TACHROME X FILM
Here Comes VonDutch

CALABASAS

Sundays with Von Dutch

Photography by Richard Karl Koch

With Text by Tony Thacker

MOTORBOOKS

First published in 2007 by Motorbooks, an imprint of MBI Publishing Company LLC, Galtier Plaza, Suite 200, 380 Jackson Street, St. Paul, MN 55101-3885 USA

Motorbooks titles are also available at discounts in bulk quantity for industrial or sales-promotional use. For details write to Special Sales Manager at MBI Publishing Company, Galtier Plaza, Suite 200, 380 Jackson Street, St. Paul, MN 55101-3885 USA.

To find out more about our books, join us online at www.motorbooks.com.

ISBN-13: 978-0-7603-2626-8
ISBN-10: 0-7603-2626-6

Library of Congress Cataloging-in-Publication Data

Koch, Richard Karl, 1934-
Sundays with Von Dutch / photography by Richard Karl Koch; with text by Tony Thacker.
p. cm.
Includes index.
ISBN-13: 978-0-7603-2626-8 (softbound w/ flaps)
ISBN-10: 0-7603-2626-6 (softbound w/ flaps) 1. Dutch, Von, 1929-1992–Pictorial works.
2. Automobiles–Decoration–Biography. I. Thacker, Tony, 1949- II. Title.
TL140.D88K63 2007
629.222'1092–dc22
[B]

2006030791

Editorial and Captioning: Dennis Pernu
Design and Layout: Tom Heffron

Printed in China

Contents

Acknowledgments

IN MAY 2006 I HAD THE PLEASURE OF ATTENDING both days of the Brucker Auction at the Petersen Automotive Museum–total immersion in a huge crowd of people who had either fond or weird memories of Von Dutch. Many of them said with their wallets what they may not have had the verbal tools to express: that the man had touched them in a way few people ever could. They wanted a bit of the dust from Comet Von Dutch.

Anyway, I hope you enjoy the book. Although I am not totally pleased with the quality of all of my original work, I am very happy with the content, and I hope you will be as well.

I feel honored and very fortunate that this book is the recipient of Tony Thacker's history in Chapters 1 through 5, and the strong coloration he gives to it all. It lends the book a character and authority it would not otherwise have.

My thanks to Mike Salisbury for the story assignment. Sheila, wherever you are, and Dutch, if you're looking down (or up) at us now, many thanks to you both for the memories.

Richard Karl Koch

west

Introduction

Von Dutch Is Still at Large!

By Richard Karl Koch

I EARNED MY FIRST MONEY FROM PHOTOGRAPHY at the University of Illinois in Urbana beginning in 1957. I had purchased a Miranda 35-millimeter SLR camera, built a wooden box around it, and took photos of the slides that were put up on a backlit panel after architectural history lectures. I then sold the photos to my classmates so they could fill in the details of the sketches we had furiously made during the lectures. Eventually,

Left: **Richard's *West* magazine cover concept, almost fully executed. Von Dutch signs the motorcycle tank of unidentified lineage in a shot similar to that which was eventually used on the May 24, 1970, cover of the magazine. Richard had been earning money as a photographer for nearly 13 years when *West* art director Mike Salisbury sent him to Calabasas to photograph Dutch, the premise being that a lot of people apparently believed the kustom culture icon was dead.**

my order for jumbo prints from the local film processor well exceeded the number of prints he made for university ID cards that year. I ordered 22,500 jumbo prints and netted a nickel profit on each back in 1959!

After going "on hiatus" from school, I landed a job as photographer for the area's leading architectural firm. Soon, I was taking similar jobs from other architects, but after about 2 1/2 years I had pretty much shot all the buildings in central Illinois and west-central Indiana, and it was time to move on–Route 66 to California in '66.

By the end of my first month in California I won first place in a photo competition sponsored by the City of Los Angeles. My prize was a Honda motorcycle. Shortly after that, I got my first cover on *West* magazine, one of the two weekly supplements to the Sunday *Los Angeles Times*. The shot was a long-lens side view of a jetliner flying low over Sepulveda Boulevard, where people gathered for the thrill of having planes thunder in for a landing less than 100 feet above their heads. But such work, while professionally and egotistically satisfying, did not pay the bills. I needed more regular work.

After trudging around town and knocking on the doors of all the top firms, my portfolio in hand, I landed assignments shooting building models for just about every major architectural firm in L.A. I shot a lot of real buildings too, but models were a whole lot more pleasant. Having seriously damaged my spine during military training in the early 1950s, my back screamed whenever I schlepped lights and gear around. I liked going to the beach on bright sunny days with puffy clouds, not dashing all over town trying to catch as many of my clients' buildings with cloudy skies behind them as I could (pre-Photoshop). With models, I could make my own clouds any time I needed them with a roll of blue seamless and a can of white spray paint.

Right: **Richard's concept sketch for the May 24, 1970, cover of *West*. Art director Mike Salisbury told Richard to "have Dutch paint a motorcycle tank with the *West* logo on it." When Richard asked Salisbury how he wanted it to look, Salisbury replied, "That's up to you." Richard left the sketch with Dutch one Sunday and by the following Sunday it was completed. Richard likes to muse that he's possibly the only person to have ever "art directed" Von Dutch. "By this time, I could have taken any rant Dutch could have given out in protest against anything," Richard recalls, "but Dutch never ranted to me at all."**

west
Von Dutch

Probably the first thing in L.A. that got me hooked photographing for fun was the proliferation of unusually painted cars, and I soon had taken a slew of photos, including Janis Joplin with her Porsche that had her band members' faces painted on it, Dean Jeffries in his shop, some Ed Roth stuff in a gallery somewhere, and lots of other wild people and objects. By that time, I had already shot two or three covers and stories for *West*, so one day I brought all my car shots to Mike Salisbury, the art director, who said, "Hmm, interesting," but not much more. Then, one or two assignments later, he called me up and said, "I want you to take pictures of Von Dutch." I said, "Who?" My photographs, I was told, would accompany Tom Nolan's 2,500-word article, which was based on the premise that many believed Dutch to be dead.

Living at the time in the low-rent part of Beverly Hills (across the street from Chasen's, a restaurant so stratospherically popular with the moneyed set that they could have charged you for smelling the food from across the street), my best-ever girlfriend, Kathy Morgan, and I packed ourselves and a few sixes of Dutch's favorite beer into *Baby Gray*, the '56 VW I had sold to my father the summer before starting college, which he later sold back to me for a buck. For a month, early every Sunday morning so we could beat the traffic, we scooted out to Calabasas to spend a day with Dutch, his wife, Sheila, their kids, and their friends and worshipers. Kinda like church, now that I think of it, but with beer, grease, and paint instead of wafers, water, and wine.

In some ways, Dutch and Sheila were so "suburban" I generally forgot that he was a living legend. The routine was pretty much the same each time. He'd already have a few pals or customers there and I think there was only one time that we arrived before he had his first beer. Once or twice he made a beer run by himself, and little by little he loosened up and in the process circled the object of his next work much as I had circled the building models I had shot as an architectural photographer, only Dutch's circles were concentric and had larger radii that included the liquor store, the streets of Calabasas, his home, his carport, and finally the vehicle. But not once in those four Sundays of copious hops-juice consumption did he ever get unpleasant, loud, or feisty—he was just an easygoing guy the whole time.

Dutch's pattern was one of nearly nonstop talk and nonstop motion. In my architectural photography life, I had the luxury of slowly circling a building looking for the perfect angle, or rotating an architectural model and playing with the lights to get an excruciatingly detailed and focused shot that often took more than an hour to set up. Now, here I was trying to catch this guy who was constantly moving and interacting with people and objects. Taking pictures of a mercurial artist was about as diametrically opposed to my usual discipline as one can imagine. This was before auto focus (or, if there was such a thing at the time, I was not aware of it and probably would not have bought it anyway). As a result, many of my photos from that era are soft. I always had trouble focusing an SLR because my right eye has an astigmatism (and now is losing vision through glaucoma), so I had to push the camera tight against my nose as I looked through the viewfinder with my left eye, a less-than-comfortable situation. I did not help matters by fitting my Nikkormat with a frosted-grid screen required for the supplementary architectural shots I was still taking.

Those four weekends were totally different from any I experienced before or since—driving from the downscale part of Beverly Hills out to the rustic charm of 1970s Calabasas and the cozy environs of Dutch and Sheila's little spread, spending the day with friendly people who were all enjoying the scene and the guy they came to see, consuming lots of beer, and having just plain good times. Sheila was an easygoing gal whom Kathy took a real liking to, and I believe the feelings were mutual because they chatted a lot all day long.

Although I knew nothing about Von Dutch until I got the assignment, my memories of those Sundays with him are very dear to me and I developed a real fondness for him. He was truly kind to me and generous with his time and my intrusions upon his idyllic life there. In fact, of all my experiences as a photographer over the span of 33 years, it was surely the most fun I ever had taking pictures. I think these photographs reflect the vibes I got from Dutch. He was nearly always smiling or laughing, except when actually working, when the

west

intensity of his concentration was palpable. It was like time suddenly stood still and as he dragged that brush tenderly across a car's skin, all eyes were lasered in on his hands. I know that mine were. When he was not striping, Dutch was mostly like a gnarly butterfly constantly flitting from one object or place to another, always studying things, working on things, moving things around, taking things apart, or putting things together.

I was really happy on May 20, 1970, the Sunday the Dutch story came out. I was celebrating one of my birthdays* a day early when Dutch called to tell me he really liked the story, giving me the best birthday present I ever got.

Richard's girlfriend at the time, Kathy Morgan, poses with the DT-1 and the finished tank. At left are "dry-run" photos Richard shot as a prelude to the final cover image. Note that in all photographs except for the image at lower right, the flames are not yet striped. That's Richard's hand at upper left, and Kathy's at lower left.

** I have two birthdays and a different birth certificate for each one to prove it. The law in Chicago's Cook County in 1934 said that a baby could not be born in a hospital without a doctor in attendance, so when my head came out just before midnight on May 21, they pushed me back in and waited for almost an hour until the doctor arrived. By the time I was "legally born," it was May 22. To this day, I absolutely cannot wait for anything and very often go bananas when I am made to.*

LOS ANGELES TIMES MAY 24, 1970
west
VonDutch '70

The *West* masthead, May 24, 1970.

west

Los Angeles Times, May 24, 1970

On The Cover

Von Dutch has striped a lot of motorcycles — even some policemen's. He says he once wore gooves in his fingers painting stripes. This one he did for WEST. Richard Koch's photo shows him signing off. More on Dutch on Page 9.

JAMES TOLAND, Director of Special Sections
MARSHALL LUMSDEN, Editor of West Magazine
MIKE SALISBURY, Art Director
UMBERTO R. TOSI, JAMES M. WIGGINS, Associate Editors
ROD KAMITSUKA, EDWARD CARBAJAL, Art Department

West Magazine, a weekly publication, is owned by the Los Angeles Times, Times Mirror Square, Los Angeles, California 90053, and is distributed exclusively each Sunday by the Los Angeles Times, which takes no responsibility for unsolicited manuscripts, photographs or other materials received. Such submissions should be accompanied by a stamped, self-addressed envelope.

Make Me A Machine That Is A Piece of Art by Tom Nolan. Back in the 1950s, a man called Von Dutch striped, flamed and painted gargoyles on countless automobiles, turning them from machines to surreal pop art creations. That could have made him wealthy and all the rage today, but it isn't his style. Actor/writer Nolan tells what Dutch—one of his early heroes—has been up to lately—which is plenty—at home in Calabasas. Photography by Richard Koch.

Von Dutch has striped a lot of motorcycles — even some policemen's. He says he once wore gooves in his fingers painting stripes. This one he did for WEST. Richard Koch's photo shows him signing off. More on Dutch on Page 9.

MAKE ME A MACHINE THAT WILL BE A PIECE OF ART

He's so legendary, some people think he's dead—which is okay with Dutch, just so they don't bother him.

By TOM NOLAN

On the thin strip of Cahuenga Boulevard that runs between Hollywood and North Hollywood an off-white building stands with the name "Dean Jeffries Automotive Styling" painted in black script. Inside in a thin fog of fiber-glass, Dean Jeffries squats beside the mock frame of a dune buggy, cutting a piece of cardboard into the shape of a fender. Jeffries has built the Green Hornet's Black Beauty and the Monkeemobile for Hollywood. Now his Kyote dune buggies are sold all over the country.

Dean Jeffries went to high school with a fellow who called himself Von Dutch. They both became professional car painters. Von Dutch earned more fame by far—he is legendary—but it is Jeffries who now has the successful dune buggy operation.

"In the old days," Dean marvels, "it was beautiful! People came from all over the country with their cars, we had 'em piled up for three, four weeks. We'd charge 'em five, six, seven hundred dollars. And they'd just say, 'Give me one of those wild paint jobs—somethin' weird—way out'."

On a far wall is a bulletin board with peeling snapshots of way-out relics—a fleet of suburban surrealist sedans adorned with giant craniums of men from Mars, razor-fanged bulldogs dripping blood, crypt-crawlers with veined eyes bulged to bursting. These jobs made Salvador Dali accessible to the man in the street.

"You don't see much of this old stuff around anymore. It's just gone. Nobody thought to save it. Hell, that was . . . 12, 15 years ago. Those cars we used to have! I remember my old '54 Merc, I had a big screaming head painted on the deck, the horn was a giant mouth, on the gas tank was a little dude droppin' in a match—now there was a *car!* People on the street would point at you when you drove that thing.

"No, Dutch was never rich. He could have been. He could have been very rich. But he'd get mad, and he'd just throw people out of the shop. Well, you can't treat customers that way if you want to have a business. But he doesn't care about any of that stuff, he never did."

When I was eight years old I knew the name Von Dutch. I knew the cars he'd painted when I spotted them in the street. I could tell them from his disciples, his style was that distinct. Ofter imitated, never equaled, thin Von Dutch stripes intersected in Brancusi spires, cool fine lines etched like spider tracings. They are my keenest memory of the 1950s.

Von Dutch lives in Calabasas now with his wife Sheila and their two little girls. His house is just off the main road and bordered by some fields. In the yard is a half-track, a World War II army vehicle with iron treads and weighing 10 tons used in desert terrain. It belongs to Steve McQueen, an old friend.

"Yes, McQueen comes here a lot," says an English parachutist who lives at the house. "It's about the only place he still can come without being bothered. It's kind of sad, really."

He was standing in his shop, a GM passenger bus, the seat-rows replaced with lathe, air compressor, metal tools and the like. He was finishing work on a circular arrangement of telegraph key switches fastened to a square of wood. "For setting off explosives in movies," he explained. "I been experimenting with acetylene. It works real good."

Von Dutch still paints cars, but the bulk of his income seems to come from other ventures. He has painted signs, hand-lettered business cards, even illustrated parts catalogues.

These days his commissions are just as varied: he completely striped a Kosher meat market in Reseda; he repairs guns; he has served as a bike wrangler and one-man special effects crew for quickie movies.

As a boy he soaked up skills; he spent after-school hours watching Simon Rodia build the Watts Towers ("Completely pure art—no function"). Now anything mechanical is as natural to him as drinking water. One of his most ambitious recent projects was the building of the Winton Flyer, the breathtaking yellow automobile of McQueen's picture *The Reivers.*

"The idea was to make a car that would embody the pleasure of motoring in the first quarter of this century, as contrasted with driving today in torture."

What he made is a dream distilled. When a tarpaulin is pulled and the Flyer /*Continued*

PHOTOGRAPHY BY RICHARD KOCH

PLYMOUTH

Von Dutch's medium is the machine. But many of his creations are lost, known—like classic Greek sculpture—only from hearsay and imitations. Unlike Praxiteles, however, Von Dutch lives and continues to work at his home in Calabasas. The photos show some of what he's been up to lately. Opposite is a flame job on an unlikely Valiant in unlikely Woodland Hills done for an engineer friend. At right and going clockwise: Von Dutch with one of his hand-wrought pistols; a magnesium tricycle with motor but no brakes that Dutch says gives a scary ride not for kids. He once made a set of motorized roller skates. Steve McQueen's WWII halftrack squats behind the trike; stripes end in spires on the back of a Mustang; McQueen's restored motorcycle sidecar; Dutch whistles to a dog from the seat of the Winton Flyer he built partly of aluminum siding for the film *The Reivers*; typically, an audience watches Dutch stripe a truck. "I think I'm going to stop painting altogether soon," he says.

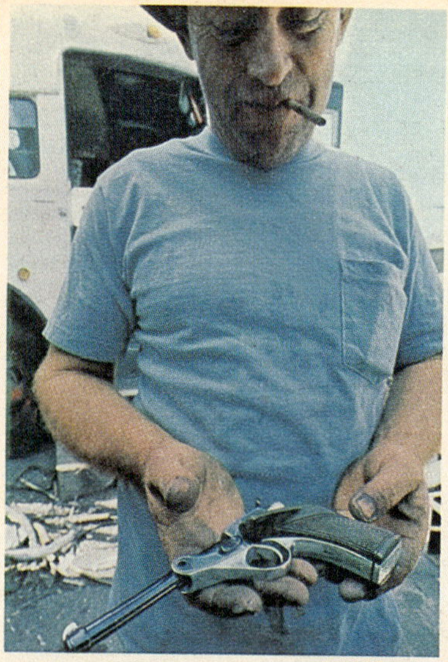

first appears on a railroad flatcar the audience receives a physical thrill.

"I worked three months building it, night and day. That was all the time I had. The body is made out of aluminum furnace sidings I found here in the yard. Some anonymous donor must have thought I could use them for something.

"So the people from the movie came out when I was finished and they said, 'Dutch, we love the car, it's beautiful, tell us, where is your shop?' And I said, 'Yer *standing* in it!' They couldn't believe I'd built the thing right here in the yard."

Von Dutch has had trouble with state regulatory agencies which discourage the operation of a one-man shop. "The air compressor I had before, they said it was too old, it wasn't safe. So I moved into the bus. Know when my new air compressor was made? 1919. And they can't do a thing about it. The bus is a movable vehicle, and their codes don't apply."

A man with an enormous gut and wearing a T-shirt arrived in a new white Mustang. He wanted Dutch to stripe it. "Bring it in out of the sun," Dutch said. He worked in short stretches, applying absolutely level lines of blue. He dipped paint onto the tiny brush every few inches, always the exact amount, for more or less would change the width of the stripe.

"Notice I lean on the car pretty heavy. And I hold the brush tight. For three years I did nothing but striping. I wore a U-groove callus in this finger from a brush that only weighs 89 grams. I striped 15,000 motorcycles and 4,000 cars. None of those are around today. They wrecked them as fast as I could paint them."

The name Von Dutch is predictably famous to hot-rod fiends; it is also known to collectors of fine modern art. As soon as Mrs. Ray Grinstein, wife of a wealthy farming-machinery manufacturer, took delivery of her new Jaguar sedan she drove to Dutch's to have it pinstriped. He finished the job in 20 minutes, once he got to it. Its spare elegance is quite at ease with the work of Jasper Johns, Al Ruppersberg, Ed Ruscha and others that stud the rooms of the Grinstein Bel Air estate. Dutch charged $15.

"We gave a party here last year for Roy Lichtenstein," Mrs. Grinstein recalls. "All the young artists came. And so did Dutch. Word spread through the house that he was here. All the artists knew who he was. 'Is *Von Dutch really* here?' one of them asked him. 'Are you Von Dutch? I thought you were *dead.*' He was the hero of the party. They all hovered around him but stayed at a respectful distance. He was a *legend.*"

"There is a mad, surrealist quality to this person," says a *Playboy* magazine editor who met him when Dutch appeared striping a naked girl in the movie, *You Are What You Eat.* "If he had been taken up and accepted by the New York art crowd, he would have received the same amount of mass-media exposure as Warhol. He preceded all those people. He made *real* pop art. I'm sure he created that whole California car ethic. He is in the Dadaist tradition."

Von Dutch admired Mrs. Grinstein's Jaguar and told her how very fond he was of that make of machine, that he had had a Jaguar himself once that was so perfect he had not even let his wife drive it. He told Mrs. Grinstein how he worked on it, painted it, raced it; and how one evening he decided there was nothing else he could do with the car, so he had a few beers, and drove it off a cliff in Topanga.

"It was finished," he explained.

Dutch denies now that he wrecked the car on purpose. "But," he says, "it *was* sort of a relief,

continued

know what I mean?"

Finished with the Mustang, he retreated into the painting shed where the sidecar from Steve McQueen's Indian motorbike sat between coats of a brilliant red. "I think I'm going to stop painting altogether soon," he said. "It is just too much of a *battle!* You're fighting the paint, you're fighting the elements, you're fighting *yourself.* I've got two kids now and I can't afford to take the time. You've got to make things *fast,* just turn 'em out."

He described an odd condition experienced at the height of fame: "I would paint a car for someone and he would bring half a dozen people to *watch me.* I realized I wasn't just painting a car, I had become an *entertainer.* To do both you have to throw yourself into some kind of dumb trance all the time." He has described perfectly the nature of being a star; but this was not a role he wished to play. "It was exhausting! I said, the hell with this."

He used some snaps of his creations to illustrate the proper painting of a car.

"It's as if something soft hits the car and is oozing back along the hood. The cars being painted today look ugly because the people painting them don't understand that everything should flow in one direction and that direction is backwards because the car is moving ahead. Now I understand how you could *see* the paint as coming forward because it's applied in reverse and your eyes might compensate."

He demonstrated this with before-and-after photographs of a flame job on a T-Bird—the before photo showed taped patches surrounded by fat vines of blood dripping down the hood; the taped patches are the flames, because it is not the flames that are spray-painted on but the spaces between them. When the tape is removed in the second photo the hood of the car becomes an optical puzzle, like the ink blot test that asks you to decide whether you are looking at a goblet or two silhouettes; with an effort of will, you discern the fire splayed by the wind back towards the driver.

"Understand? The car is moving forward so everything on it flows back. That way it's part of the natural river, it's in harmony with nature. Like a tree, you never see one thicker on top than the bottom, all those branches taper off. It's one of nature's happenings."

Some years ago, Von Dutch devoted his time to the manufacture of exquisite firearms, replicas of dueling pistols and antique rifles. He aged the metal. He aged the velvet in the pistol cases. Expert collectors thought these guns the genuine article.

All the guns were made completely by hand — no metal was cast. Von Dutch hollowed lengths of iron with his tools, "hogged the guts out. That's the fun of it, doing it yourself." All the guns, both breech and muzzle-loaders, fire perfectly. Gunsmithing is just another master skill Von Dutch acquired in his youth. The consummate beauty of these arms results from the rich engraving Von Dutch applied along the barrels.

The engraving is done with a thin little tool he made, no more than a sharp piece of metal. "Engraving takes an enormous amount of strength. I don't like to do it for gun collectors. The only thing they give a damn about is that the engraving should be exactly the same on both sides of the barrel. It gets boring."

That uncompromising streak in Von Dutch has marked his career. "Someone once explained to me why I could never have a factory that turned out motorbikes. 'Because you would always be in there tearing the bikes apart. You'd never be satisfied because you'd know no assembly line could make the bike as well as you do, the right way, just building up from the ground'."

It was evening now. He walked through the yard past some disassembled slot machines. "A fellow brought some machine sculpture he'd done out here he wanted me to take a look at.

"He had some pretty cute things. But I like a machine to *do* something, even if it's only making popcorn."

Von Dutch made two remarks that seem to be related.

Once, after a technical speed-speech about an energy-transfer system — Dutch talks sometimes like Zen masters who realize only a small part of their audience comprehend them but who refuse to speak on a lower level — he said, *"Obsolete* doesn't mean a thing isn't any *good*—it just means they don't *make* it anymore."

Later when it was dark and Dutch sat slumped in a chair in his living room talking about Bugatti, the man he believes was the last real artist of car design. He told how Bugatti's factory made every single part for his automobiles, even the door locks.

He said Bugatti returned mechanisms that were not really reliable, like the brake system, an intricate and erratic network of chains and gears. When his cars were criticized because of the brakes, Bugatti said, "I don't make them to stop. I make them to go."

Now he waved vaguely at the stars above. He said, "Every time someone I don't know shows up here it's heavy over my head—I'm afraid it's somebody come to make me close up shop. I don't think it's a bad thing that a man can live by working alone and making things with his hands. I'm afraid I'll have to go back repairing motorcycles in some garage."

Von Dutch may never have security—it seems alien to the way he has always worked, creating masterpieces from discarded aluminum siding. Von Dutch may never be rich; he will not waste much time crying over that. As for me, I am happy to say I have met a genius.

I am going to give Dutch the money I earn from this article, and he will make me a machine that will be a piece of art. And it will work, too.

Chapter 1

Calabasas, 1970

IN EARLY 1970, THE NATION WAS IN TRANSITION. Detroit was riding high, pumping iron at the rate of 6.5 million units per year, with Cadillac offering the world's largest-capacity engine: 500 cubic inches, more than four times bigger than the average Honda. Chrysler's Hemi 426 was the world's most powerful production automobile engine with 425 horsepower. The Plymouth Hemi Super Bird Coupe, a full-blown, street-legal NASCAR race car, could be had for $3,600.

Left: **Richard points out that Von Dutch had not only his famous GM bus/machine shop in Calabasas, but also a small wooden shed. "But you had to be careful to not call it a shed in his presence," Richard says. "It seemed that the City of Calabasas had zoning restrictions prohibiting outbuildings. To avoid problems with the bureaucrats, Dutch attached some long planks to the bottom of the shed and, *voila*, a sled, which was, of course, OK to have on property in Calabasas, even though it never snows there."**

The feature on Jack Chrisman's Dutch-striped '29 Tudor in the December 1954 issue of *Hot Rod* was followed by more ink a few months later when the March 1955 *Rod & Custom* featured George Barris' heavily Dutched '49 Ford woody on the cover **(see page 26).**

• No, that radiator didn't spring a leak—just some inspired striping by famed southland, brush wielder, Von Dutch. '35 Plymouth front axle is fitted with '40 Ford spindles; '46 Merc brakes, Columbus shocks. Car turned a record 114.64 mph.

• '29 Model A body is gutted and sprayed with underseal inside. Jack's famed nephew, Art, did most of the engine work.

• Front tires are 6.40 x 15s. Rear slicks, 8.20 x 15s, were cast in a special mold, capped with soft gum rubber and have 8½ inch width. No radiator is used; water retained in engine.

• 276-inch Merc used Harman and Collins mag. Sharp heads and manifold. Art Chrisman worked heads over to handle mixture of 75% nitro, 25% alcohol. Speed secret was in carburetion.

Photos by Rick

• 5½-inch chopped top was covered in white by Russ and Bess Top Shop. Body is sprayed with maroon Ditzler lacquer. Rear has Houdaille shocks put ahead of axle.

TORRID '29 TUDOR

Another Chrisman Screamer Hits the Strip

When the Chrisman clan launches a drag job there's never any question on two points—appearance and speed. A recent bomb to emerge from the famed Compton, California, garage is a torrid 2200-pound tudor owned by Jack Chrisman and built with the assistance of his famed nephews, Art and Lloyd.

Chassis modifications include new crossmembers front and rear to accommodate the '46 Merc engine, '39 Merc transmission and '40 Ford rear end (4.11 gears). After reversing the eyes of the front spring, it was found necessary to put a little more arch in the leaves to set the car level. If you are an old Plymouth fan, you may recognize that front axle, a refugee from an ex-bellytank lakester. The rear spring setup is strictly stock and handles the load very well.

Sharp equipment dominates the outside of the engine; the manifold fitted with four "97" Strombergs jetted for fuel. The inside job is handled by a Potvin ⅜ cam and Forged True pistons pumping 10½ to 1 compression. Displacement is 276 cubic inches. Actually, this bit about the engine should be phrased in the past tense since the little mill scattered itself several weeks ago after posting a top speed in the "B" Fuel Class at Pomona at 114.64 mph.

And if the record isn't enough to scare off the competition, that wild Von Dutch striping job should shake 'em up a little. Bought originally from a junk yard for $15, the A Model was scheduled to be a street job at first. But, as Jack points out, who'd have the heart to send a little jewel like that out into the brutal Los Angeles freeway traffic?

...nd '49 Pontiac taillights dress up tail and drag strips after owner 'chickened out' on city streets . . . can you blame him?

• Dash was taken from a '39 Ford, fitted with a complete selection of Stewart-Warner instruments. Late steering used.

HOT ROD MAGAZINE

• Fuel tank and seat sit side by side. Seat is type used in small English 'Theme' (Ford). Fuel filter, fuel tank, shut-off valve and fire extinguisher are all aircraft surplus items.

DECEMBER 1954

• Gettin' under way is the thing Chrisman does best. Noticing a distinct twist in frame. Jack braced frame with a 2-inch tube under the engine. In eight times out, the car won four trophies.

27

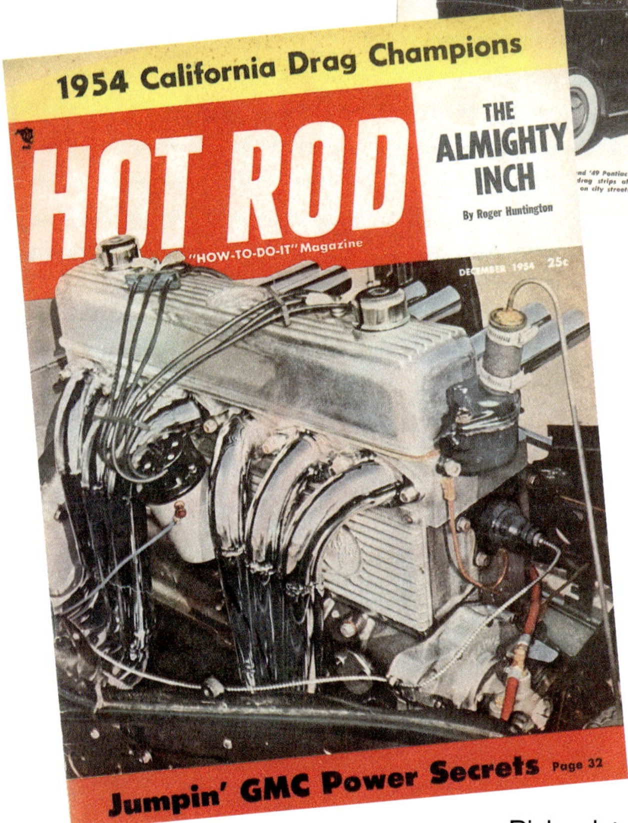

Not everything was roses, though.

A few months earlier, the Rolling Stones' "free concert" at Altamont Speedway in Northern California ended in a crescendo of violence that many still consider the official death knell for the '60s. Boys who had turned to men in the jungles of Vietnam were arriving home from the first war that didn't send back heroes.

Richard Karl Koch was hugely affected by these events. One day in late 1966 over lunch in the University of Illinois student cafeteria (although Richard was by then a full-time photographer, he still maintained old college friendships), he suggested to his college pal, Roger Ebert, that if all the GIs in Vietnam stood 15 feet apart, they could walk north, shoot everything in sight, and end the war. Roger told him he needed an education. He took Richard to his apartment and dug out some reading material. Richard was stunned by Vietnam's history. He wrote what he titled "A White Paper on the U.S. Role in Vietnam," and circulated it to everyone he knew, including each

During his Sunday visits to Calabasas, Richard photographed Von Dutch's business card. Judging by the amount of press Dutch received in 1950s enthusiast magazines, his claim to be the (or simply *an*, depending on how you read it) "Innovator of Modern Automobile Pinstriping" was no exaggeration. The use of the word "Modern" is also interesting, perhaps a specific reference to the postwar period, by which time manufacturers had abandoned striping in favor of chrome trim, leaving the art largely to stripers working as freelancers or out of automotive dealer lots.

client of his flourishing new architectural photography business. Within three weeks Richard, who had been injured while training for the Korean War, was completely out of business, having been labeled a "communist," and worse. It was the middle of a harsh Midwest winter and "California Dreamin'" was atop the charts, so Richard headed out to La La Land.

Disillusioned by the hard line of his former clients and chafing from the mercilessly unforgiving constrictions of architectural photography, Richard sought subject matter that represented freedom from constraints. In early 1970 he had wrapped up a major *Mankind Magazine* photo essay on the relatively young Renaissance Pleasure Faire, and a story for *L.A. Times' Home Magazine* about a young couple living on their antique-filled, nineteenth-century Baltic Sea trading schooner. He had also put together a sizeable portfolio of car photography that he brought to Mike Salisbury, the art director of *West* magazine who had already given Richard several freelance assignments for the *L.A. Times*' Sunday supplement.

> Before causing a sensation at *West*, Salisbury had designed a surf magazine. Later, he would art-direct *Rolling Stone*, oversee ad campaigns for hundreds of films, and earn a degree of notoriety for his creation of cigarette icon Joe Camel. In the spring of 1970, he suggested that Richard and writer Tom Nolan produce a story on Von Dutch.

The quick succession of Jack Chrisman's drag-racing '29 Tudor in the December 1954 issue of *Hot Rod* (see page 24) and this March 1955 cover of *Rod & Custom* featuring George Barris' '49 Ford woody did perhaps more than anything to put Dutch's style of "modern pinstriping" on the map.

Dutch had been working for actor Steve McQueen for a couple of years, on the car for the 1969 film *The Reivers*, and at the Le Mans race circuit in France. According to a sign that Dutch painted, and which hung on a wall in Valley Ford Parts in 1970, he was "Glad to be back home." Home, Richard recalls, was a ranch house, Dutch's famous machine shop/GM bus, and a workshop on runners (the runners having converted the workshop from shed to sled, against which there were no zoning laws in Calabasas).

Legend has it Kenneth "Von Dutch" Howard started his career at various motorcycle shops around the Lynwood, South Gate, and Compton areas south of downtown Los Angeles: Art Tatum's, George Beerup's, and Shell Tuitt's, to name a few. Back in the late 1940s and early 1950s, this area of L.A. was ground zero for the exploding hot rod and custom car scene. However, it would be some years before owners of the four-wheelers adopted pinstriping as a means of decorating their rides, whereas motorcycle tanks and fenders were still being hand-striped at factories well into the 1960s and with the arrival of Japanese imports.

Automobiles had not been factory-striped since before World War II and the introduction of chrome and stainless-steel trim. They were, however, being striped at dealers' lots by guys like Tommy "The Greek" Hrones and Sal Guerrero in the Bay Area, and by Art Summers and Bud "The Baron" Crozier in the Los Angeles region. The Baron's business card, circa 1958, read, "50 years experience pinstriping."

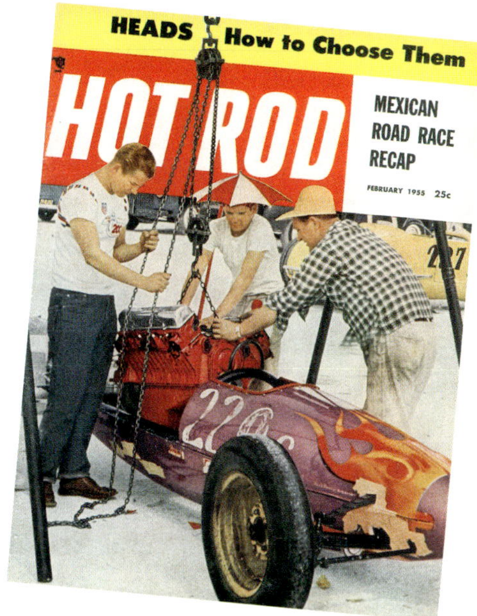

Perhaps what put automobile and, indeed, Dutch's style of "modern" pinstriping on the map was the appearance of Jack Chrisman's drag-racing '29 Tudor in the December 1954 issue of *Hot Rod* magazine and the cover feature on George Barris '49 Ford woody on the March 1955 issue of *Rod & Custom*.

The Reed Bros./Leroy Neumayer belly-tank lakester that appeared on the cover and inside the February 1955 issue of *Hot Rod* provides early photographic evidence of Von Dutch's trademark flying eyeball. No doubt, this issue of the magazine, which also included Bob D'Olivo's feature "Camel's Hair Capers," helped further catapult Kenneth Howard into the limelight. The striping boom exploded later that year.

Winton
Flyer
CALIFORNIA
HORSELESS
CARRIAGE
1119

Prior to that, neither customs nor hot rods were striped much; the only vehicles that were striped were pickup trucks (along the belt line and around the reveal lines in the bed sides, etc.). Also, what flame and scallop jobs there were back then went, for the most part, without outlines.

Evidence of early pinstriping is seen on Bob Brown's '40 Ford Coupe featured in the February 1954 issue of *Rod & Custom*. Sadly, no credit appears for the wild, multicolored flame job, but it seems to have some edge, which may have inspired others. Even earlier examples include Dick Kraft's track T that had outlined scallops in 1952, and Frank Rose's Bay Area–based '27 T seen in the June 1953 *Hot Rod* with Tommy the Greek striping around the panels and louvers. Rose's T may well have been one of the first deviations from traditional striping, but it was tame compared to what Dutch was concurrently doing down south.

According to *The Best of Hot Rod* (Petersen Publishing, 1981), edited by Bruce Caldwell, in 1954 *Hot Rod* magazine put Dutch in a booth at the International Motor Revue and Motorama and had him stripe a '34 tub. Unfortunately, no photographs of the event appear to have survived and it's not even clear whether Dutch ever did stripe at the show.

Left: **Von Dutch in the Winton Flyer, a witch's brew of parts he crafted into the car that costarred in the 1969 Steve McQueen film *The Reivers*. Dutch's old friend Phil Lebow later reminded Richard that, in this photograph, Dutch was whistling in an attempt to coax a stray dog to urinate on the wheel, thinking it would make for a more compelling shot.**

Above: **Fifteen years after the initial pinstriping boom, Von Dutch was still hard at it when Richard visited his home in Calabasas. Richard recalls, and his photographs verify, that Dutch's skill could still attract throngs of admiring fans.** *Right:* **Pinstriping was just one of Dutch's fortes, of course. Richard also captured him at work in his bus/machine shop. The gun was reportedly commissioned by Tom Nolan, who wrote the *West* article that accompanied Richard's photography.**

However, it is known that, on display in the Pan Pacific Auditorium was the Reed Bros./Leroy Neumayer belly-tank lakester and the Chrisman Bros. coupe, both of which were heavily "Dutched." In fact, photographic evidence shows that both also displayed Dutch's trademark flying eyeball: the belly tank on the nose and the coupe on the rear splash apron.

No doubt, all this publicity and the later Bob D'Olivo feature "Camel's Hair Capers" and the Bob Greene/Eric Rickman cover story on the Reed Bros./ Neumayer tank—both in the February 1955 issue of *Hot Rod*—as well as the subsequent Rickman photos in the February 1956 *Car Craft* of Dutch with striping brushes in his ears, catapulted Kenneth Howard and his style into the limelight.

The striping boom exploded in 1955, and magazines began to feature stripers, such as Art Summers (*Rod & Custom*, March 1955), and striping techniques.

Dutch's first love, however, was motorcycles, and he never appeared to fully embrace the car culture. Consequently, when it came time to execute Salisbury's and Richard's concept for the cover of *West*, Dutch used a castoff tank mounted on single-cylinder, two-stroke Yamaha DT-1 dirt bike as his canvas. Richard dropped the sketch off one Sunday and it was done the next. "By this time," Richard says, "I could have taken any rant Dutch could have given out in protest against anything, but Dutch never ranted to me at all."

Above: **Richard and his girlfriend, Kathy Morgan, who accompanied him to Calabasas, spent time in Dutch's and Sheila's home, as well. Dutch created this lamp from an old dynamite plunger box.** *Right:* **Outside the home and on area roadways, there could be no mistake as to the owner of this Corvette-powered Chevrolet shop truck... especially for those looking in their rearview mirrors.**

Here Comes Von Dutch
CHEVROLET

The view from here
The weld and the rock

Chapter 2

Cool Fine Lines

WHILE VON DUTCH HONED HIS CHOPS STRIPING MOTORCYCLES, the curvilinear canvas of automobiles propelled him into the spotlight. A striper needs little in the way of equipment. Doesn't even need a shop, really. Many stripers both then and now travel from job to job with all the tools of their trade in a small box. A few dagger brushes and a couple of tins of 1-Shot are all it takes.

Left: **"Dutch had this little three-legged table upon which he put a magazine and a small round paint container," Richard recalls. "Then he dipped the brush into the paint, swished it around, lifted it out, and lightly jiggled the long bristles back in alignment and started the stripe." Richard nicely captured the entire sequence of Dutch striping a new Mustang.**

Right: **It's fair to say that, today, most think of automotive pinstriping as a conceit of hot rod and custom enthusiasts alone. In reality, the form predates automobiles and has historically known no automotive boundaries, national or otherwise. Von Dutch striped this Jaguar (see also pages 54–55) for L.A.-area arts patron Elyse Grinstein, from whom he extracted $15 for his services. Nolan wrote that the job took Dutch 20 minutes.**

Take, for example, the photos in which Richard Koch captured Dutch striping a new, white Mustang. Dutch mixed his paint in a shallow tin and then carefully loaded his brush on any handy piece of paper. As he stated in a February 1956 interview in *Car Craft* magazine, "The secret of striping is loading the brush right, not the striper."

Every striper has a distinct technique. Dutch's is described as "pinkie down, two handed," which means he held the brush in his right hand and steadied his brush hand with his right small finger (his pinkie) as well as with his left hand, hence the two-handed moniker. The first published photograph of Dutch striping appears to be on page 32 of the February 1955 issue of *Hot Rod* magazine as the lead shot for a feature entitled "Camel's Hair Capers." The opening sentence quotes Dutch: "I call them amoeba designs because they are constantly changing their shape as I go along, bending to the forces of environment and mood."

The writer (while Bob D'Olivo was credited with the photographs, no credit was given to the writer, though it was most likely also D'Olivo) went on to state, "A Von Dutch original [is] an absolute must for many a southland hot rodder and custom fan."

"Camel's Hair Capers" was followed the next month by Spence Murray's six-page story about striper Art Summers in *Rod & Custom*. Suddenly, striping was the big trend. It was quick and visually stunning, and soon Dutch and Summers, who were reputed to be good friends, were not the only artists who plied their trade with a dagger.

The cover of that March 1955 issue of *Rod & Custom* featured George Barris' shop truck, a "Dutched" '49 Ford wagon. There was nothing about the truck in that issue, but the February 1956 *Car Craft* had a 7 1/2-page feature on Dutch entitled "That's Me." Dutch was interviewed by Jack Baldwin, who took pains to explain Dutch, who seemed to revel in obscure esoteric answers and whose most profound statement was, "The biggest secret [of striping] is to take your mistakes and make it look like you did it on purpose."

Pinstriping not only transcends automotive boundaries, it transcends automobiles, too. Richard also photographed the stripes Dutch laid on the cash register, bank vault, pay phone, and soffits at Bill Fredrick's butcher shop near Canoga Park. The pinstriping certainly worked with the period décor at Fredrick's. Pinstriping had been widely practiced during the Victorian Era, until about the end of the twentieth century. In his book, *How to Pinstripe* (Motorbooks, 2007), Alan Johnson notes, "A variety of ornamentation and decoration was painted onto each and every machine that existed during this time." More examples of Dutch's striping at Fredrick's can be seen on pages 58–59.

The feature ended with a full-page, step-by-step photo essay of Dutch creating his infamous Rabbit: "Harvey–shaken by cross breeding." The art and his antics in the accompanying photographs only enhanced Dutch's reputation to a generation demanding more expression and individuality. Reinforcing his position as the don of modern design was the explanation of one of the panels on Barris' wagon: "The painting tells the story of a man who eats a bowl of chop suey while walking through a valley of mountains contemplating a terrible suicide plot." If, as Barris says, Dutch was locked in his shop and drank all night while he striped the wagon, then perhaps the explanation is not so farfetched. Or was Dutch just having fun with us?

Two months later, for April 1956, George Cerney's '50 Plymouth two-door wagon was featured on the cover of *Car Craft*, and again Dutch was credited thus: "Von Dutch, So. Calif. striping king, detailed the wagon with eccentric designs." Like it or not, Dutch was well and truly in the limelight–a pop artist even.

Beyond the automotive magazines, Dutch began to appear in popular culture outlets. Magazines, such as *DIG,* featured him under the title, "This Guy is Crazy." In the June 7, 1958, edition, the *Los Angeles Herald & Express* said of him, "This off-beat crew-cut character is rapidly acquiring a reputation as the Salvador Dali of automotive painting." Other articles followed. He was invited to artsy parties and a reluctant star was born. Elyse Grinstein, who had the Jaguar seen here striped by Dutch for $15, told Tom Nolan, "We gave a party here last year for Roy Lichtenstein. All the young artists came. And so did Von Dutch. Word spread through the house that he was here. All the artists knew who he was. 'Is Von Dutch really here?' One of them asked, 'Are you Von Dutch?' He was the hero of the party. . . . He was a *legend*."

(Interestingly, Richard points out that while Tom Nolan identified Mrs. Grinstein as the wife of "wealthy farming-machinery manufacturer" Ray Grinstein, the Mr. Grinstein who hosted the Lichtenstein party was actually Stanley Grinstein, a prominent Los Angeles–area arts patron.)

The *West* article goes on to quote a *Playboy* magazine editor: "There is a mad surrealist quality to this person. If he had been taken up and accepted by the New York art crowd, he would have received the same amount of mass-media exposure as Warhol.

Among the many admirers and friends who gathered in Calabasas on those four Sundays in early 1970 was Norm Grabowski, creator of the *Kookie Kar* T-bucket and an icon of kustom culture in his own right. Dutch honed his pinstriping chops on motorcycles, and here Richard captured his reflection–hands on hips–in the tank of Grabowski's Corvair-powered motorcycle, known as *Six-Pack*.

In the *West* article, Tom Nolan reported that "A man with an enormous gut and wearing a T-shirt arrived in a new white Mustang. He wanted Dutch to stripe it. 'Bring it in out of the sun,' Dutch said." Richard captured Dutch assessing the situation shortly after the car's arrival. Dutch told Nolan that he had striped 4,000 cars in his day, in the process wearing a U-shaped callus into his right middle finger.

Dutch prepares the Mustang's surfaces (right) before beginning the job (opposite). "He might start with an elaborate series of stripes, points, and arcs before heading back along the car, or he might just make a simple arc or flourish, or sometimes nothing but a stripe before heading back along the car," Richard remembers.

He preceded all these people. He made real pop art. I'm sure he created the whole California car ethic. He is in the Dadaist tradition."

In addition to the Grinstein Jaguar—a 1967 or 1968 Mark II, judging by its slim-line bumpers and license tags—another Jag featured prominently in Dutch's life. Around 1957 he owned a powder-blue XK120. In the few photographs that remain of the car, it appears customized with frenched headlights and a custom hood with a scoop and louvers. Interestingly, it looks very similar to a '51 XK120 that George Barris owned and which also had a similarly scooped hood and frenched headlights. Similarities notwithstanding, Dutch claimed to have wrecked the car one night on Mulholland Drive while going home from a jazz club. As Nolan wrote of the Jag in *West*, "He told Mrs. Grinstein how he worked on it, painted it, raced it; and how one evening he decided there was nothing else he could do with the car, so he had a few beers, and drove it off a cliff in Topanga."

It was a story Dutch later retracted.

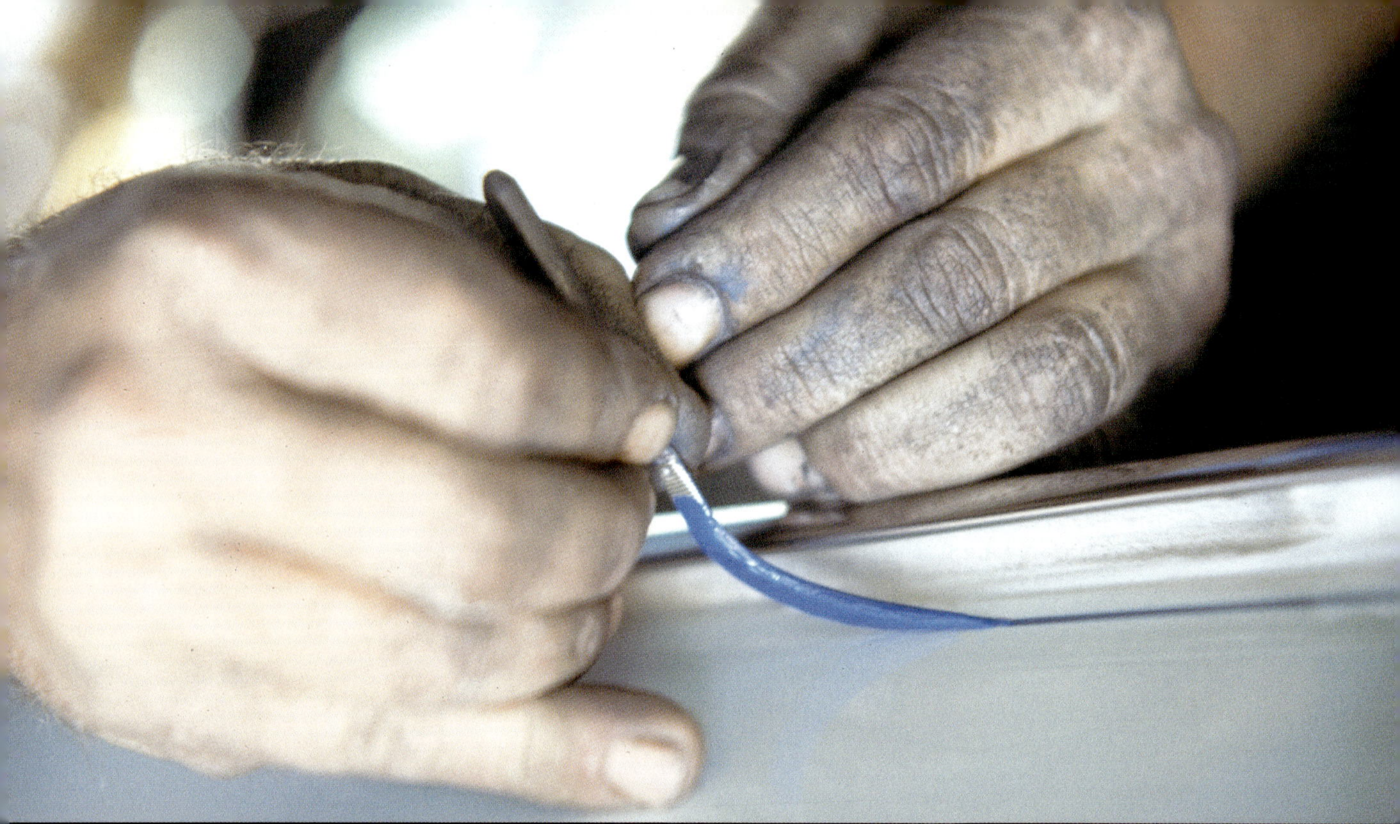

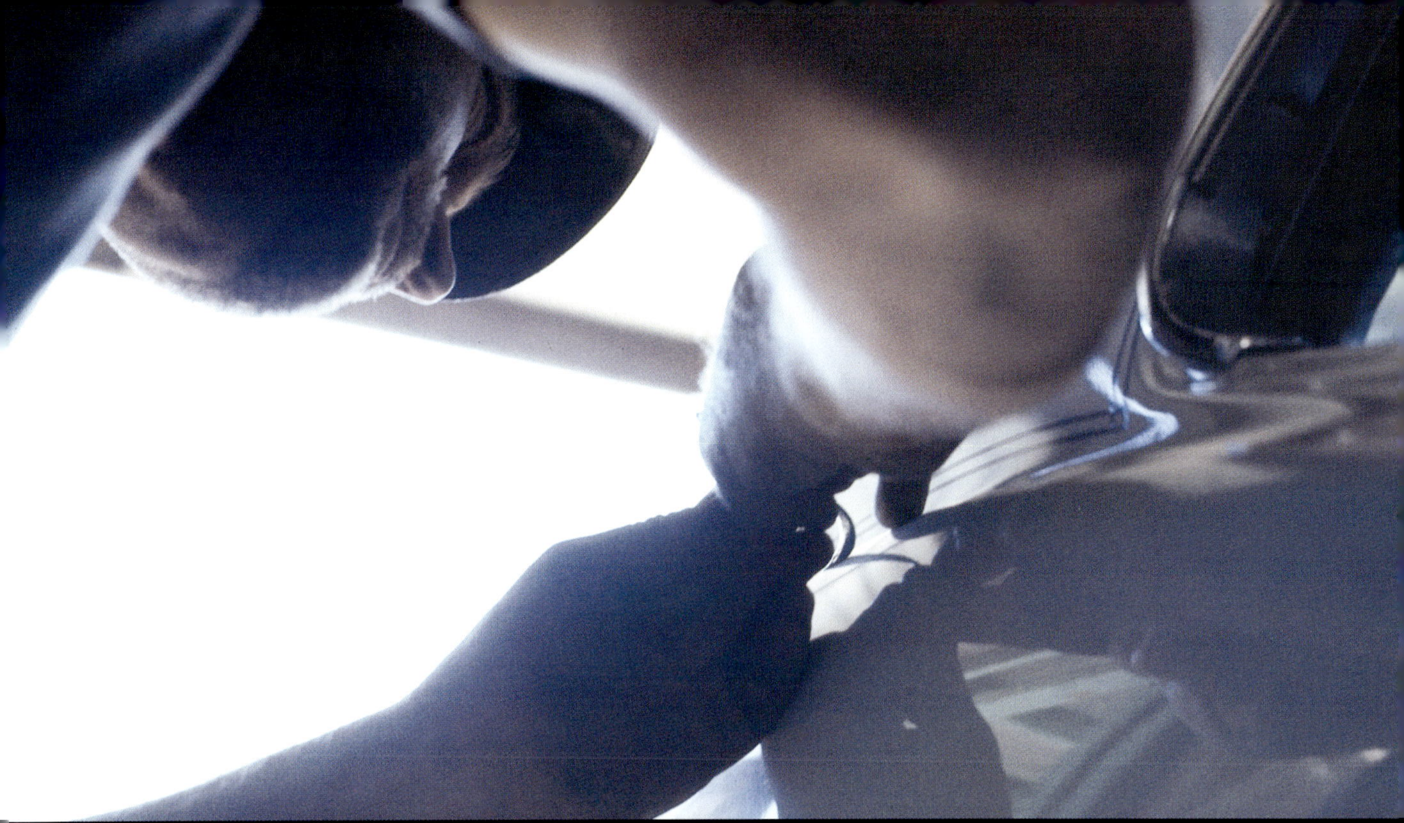

Every striper has a distinct if not necessarily unique method. Dutch's is described as "pinkie down, two handed," which means he held the brush in his right hand, steadying it with his right small finger (his pinkie) and with his left hand. Richard captured the technique brilliantly in the sequence on this page. As for his own striping philosophy, in 1956 Dutch told Jack Baldwin, writing for *Car Craft* magazine, "The secret of striping is loading the brush right, not the striper."

“Dutch was able to draw a heckuva long line before having to reload the brush,” Richard says, “and it was amazing to watch him lift the brush from the line and then return to the same spot and continue on without the slightest indication that the line had been interrupted and restarted.” Nolan claimed that on the Mustang, Dutch “dipped paint onto the tiny brush every few inches, always the exact amount.”

Of the Mustang job, Richard remarks, "Eventually, he was at the trunk lock, where he laid down more swirls or arcs, and that was it for one side of the car. Then, he'd do the other side the same way. I was unable to see any variation between the two sides most of the time, but I am told sometimes he would intentionally make one side slightly different than the other."

The Ford

The Ford

Another vehicle that Richard documented Von Dutch pinstriping was L.A. Roadster member Carl Riggen's rodded '29 Roadster pickup painted Ford "Peacock Blue." Riggen, who was a sound engineer at MGM Studios, is seen sitting on the cinderblock in the top two photographs. That's writer Tom Nolan, wearing the striped T-shirt in the lower photos.

Top: **"The routine was pretty much the same each time," Richard explains of his visits to Calabasas. "Dutch would already have a few pals or customers there and I think there was only one time that we arrived before he had his first beer." Richard remembers picking up a couple six packs of Dutch's favorite beer on the way up to Calabasas each morning. Some of the pals in this photograph include (from right) the Model A's owner, Carl Riggen; Phil Lebow (moustache); and Norm Grabowski (flying helmet).** *Right:* **Lebow sports a T-shirt with a silkscreened image inspired by a photo that appeared in *DIG* magazine 14 years prior. Perhaps Lebow's T-shirt can be considered the forerunner to a designer clothing empire launched a quarter of a century later.**

KODAK
on Dutch
IS STILL AT
LARGE!
34A

JAGUAR
CALIFORNIA
TVE 027

CALIFORNIA
TVE 027

This Plymouth Duster couldn't have been long from the factory when its owner took it to Dutch for the flames and pinstripes seen here. The Valiant-based Duster's first model year was 1970 (released in late '69, the date next to Dutch's signature), and this 340 was the high-performance version, rated at 275 horsepower. Richard's creative composition on the facing page shows just how much a Dutched car stood out from the crowd. The stripes on the rear deck lid are those which Richard projected onto Dutch for the photograph on page 7.

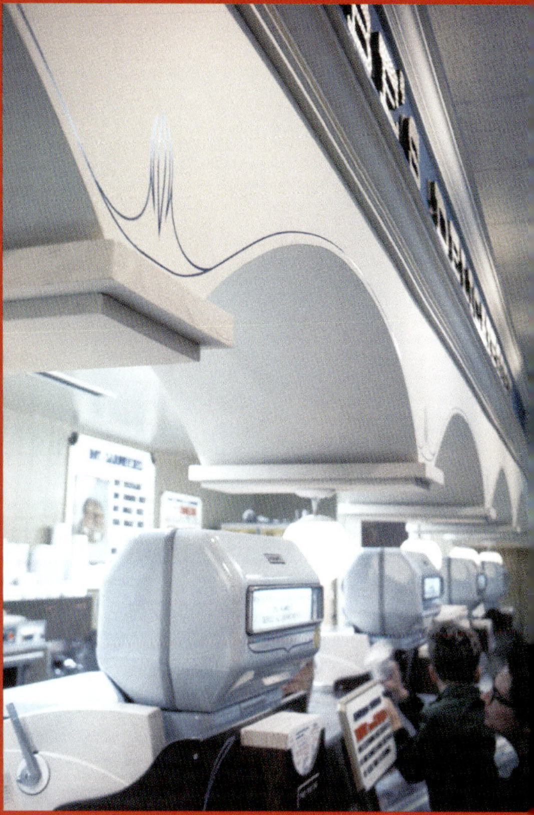

More detail photographs of Von Dutch's pinstriping at Fredrick's butcher shop near Canoga Park. In the story he penned about Richard for issue No. 31 of *The Rodder's Journal*, noted car designer Thom Taylor points out that Bill Fredrick was himself a hot rodder, whose slingshot-style T-bucket appeared in numerous magazines throughout the '60s, and who also had a hand in a number of land-speed vehicles.

Richard ventured out nearer his home, to Sunset Boulevard in Hollywood, to photograph this Mack that Dutch painted and striped in 1969. The work on the door harkens back to Dutch's roots–his father was a sign painter and Dutch himself would more extensively practice the craft later at Brucker's Movie World. On its marquee, the landmark Cinerama Dome in the background advertises the Clint Eastwood vehicle *Paint Your Wagon*. Richard says that he asked the owner to move the truck so that he could get the marquee into the shot. "I was always paying attention to billboards and signs and how they appeared in certain contexts," he explains. "Gary Krueger, my longtime friend and a fantastic photographer, may have influenced this attention in me." Richard also remembers that he came across the rig purely by chance and asked the owner if it had been "Dutched."

Von Dutch's gearhead roots were in motorcycles (he told Nolan he had pinstriped 15,000 of them), and there seems never to have been a shortage of them at Calabasas when Richard visited. As with the Mustang and Duster owners, two-wheel aficionados were eager to bring the latest and greatest to Dutch for a bit of personalization. Above, Dutch and the owner of the white Mustang previously shown examine a Honda CB750. Quite simply one of the most influential motorcycles ever, the four-cylinder inline was launched in 1969 and has been credited widely for changing the course of motorcycle history. This example exhibits some comparatively mild Dutching. The helmet on the other hand.... Judging by its shape, the tank at left is possibly an aftermarket item.

This page: **Dutch applies an interesting motif to a Triumph tank and then holds court with writer Tom Nolan. If legends are true, flames and stripes weren't his only stock-in-trade. Richard recalls a story he was told in which a customer insisted on a particular pinstripe design for his bike tank, but which Dutch insisted wouldn't work. The customer pressed, however, and Dutch relented . . . sort of. The tale's payoff, as related to Richard by Phil Lebow, was that Dutch acquiesced, but the "stripes" comprised a continuous string of "FUCK YOU" hand-painted in miniscule capital letters. If it sounds improbable, refer to the photograph on page 124.** *Opposite:* **Another new bike that appeared in Calabasas and which Richard documented was this 1970 Kawasaki Mach III. Note how well Dutch's work complements the tank's distinctive scallop–a far sight better than the horizontal racing stripes that adorned stock models.**

Chapter 3

On the Bus

"VON DUTCH'S MEDIUM IS THE MACHINE," Tom Nolan wrote in *West* magazine, but where did Dutch get his mechanical affinity? It's one thing to be able to pinstripe a motorcycle or a car, but it requires a whole other skill set to fashion guns, knives, and machines from little more than scrap metal and wood. Of course, Dutch was self-taught in everything he did, from playing the flute to turning on a lathe he had fixed himself. In *The Art of Von Dutch* (Tornado Design, 2006), Suzanne Williams recalls, "In fact, he was

Left: **A fine example Von Dutch's handiwork–and his hands. Richard recalls frequently being mistaken for Steve McQueen on the streets of Hollywood in the 1960s. Of first meeting McQueen, Richard recalls, "As we shook hands I said, 'Steve, I promised myself that if I ever met you I would ask you, "Did anyone ever tell you that you look like Richard Koch?"' Completely missing the irony, he replied by showing me his palms and saying something like, 'Do your hands look like mine?' I later learned that both he and Dutch judged men by their hands. Of course, I surely must have come up short."**

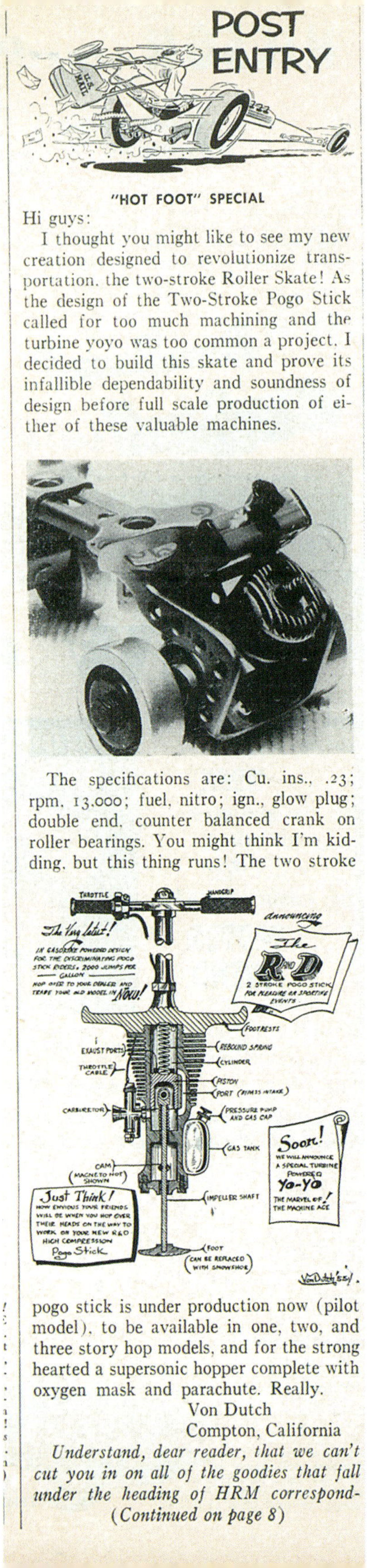

POST ENTRY

"HOT FOOT" SPECIAL

Hi guys:

I thought you might like to see my new creation designed to revolutionize transportation, the two-stroke Roller Skate! As the design of the Two-Stroke Pogo Stick called for too much machining and the turbine yoyo was too common a project, I decided to build this skate and prove its infallible dependability and soundness of design before full scale production of either of these valuable machines.

The specifications are: Cu. ins., .23; rpm, 13,000; fuel, nitro; ign., glow plug; double end, counter balanced crank on roller bearings. You might think I'm kidding, but this thing runs! The two stroke

pogo stick is under production now (pilot model), to be available in one, two, and three story hop models, and for the strong hearted a supersonic hopper complete with oxygen mask and parachute. Really.

Von Dutch
Compton, California

Understand, dear reader, that we can't cut you in on all of the goodies that fall under the heading of HRM correspond-

(Continued on page 8)

so completely involved in obsolete motorcycles that he could machine many of the parts from memory."

Early photos of Dutch in his apartment that appeared in the March 1956 issue of *DIG* magazine show several lamps fashioned from musical instruments—a bugle, a trumpet, and even a saxophone—each adorned with a Dutch-made and decorated lampshade. This was in the years before mass consumerism, when kids still improvised their toys and nobody gained their experiences by watching TV. Some of those early *DIG* shots also show homemade cannons and roller skates powered by model-airplane engines. Then there were the engineering drawings of the infamous motorized "Two-Stroke Pogo Stick" that appeared in the letters section of the November 1955 issue of *Hot Rod*. Dutch's letter also advertised "two-stroke Roller Skate" [sic] and mentioned a "turbine yoyo."

The letter only added to Dutch's zany reputation. *Hot Rod* editor Wally Parks replied, "Understand, dear reader, that we can't cut you in all of the goodies that fall under the heading of HRM correspondence, but occasionally your staff just can't resist flanging some of the not-so-conventional material." In the following June's issue of *Car Craft*, a pogo-riding Dutch appeared in a story titled "The Legend of Ledhed McSlabb," written and illustrated by Carl McKohler.

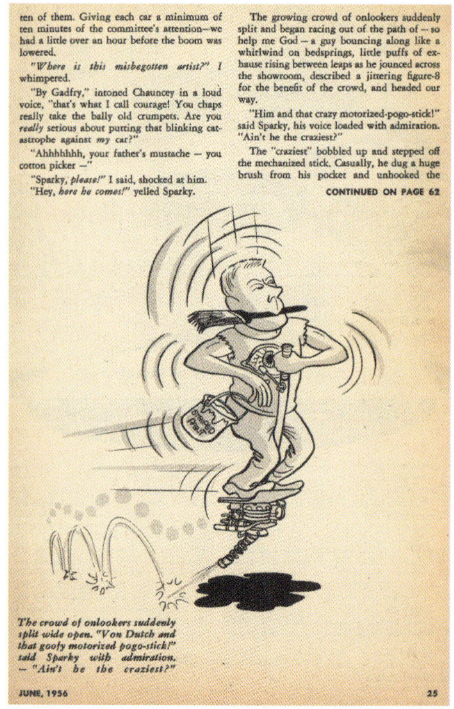

ten of them. Giving each car a minimum of ten minutes of the committee's attention—we had a little over an hour before the boom was lowered.

"Where is this misbegotten artist?" I whimpered.

"By Gadfry," intoned Chauncey in a loud voice, "that's what I call courage! You chaps really take the bally old crumpets. Are you *really* serious about putting that blinking catastrophe against *my* car?"

"Ahhhhhhh, your father's mustache – you cotton picker –"

"Sparky, *please!*" I said, shocked at him.

"Hey, *here he comes!*" yelled Sparky.

The growing crowd of onlookers suddenly split and began racing out of the path of – so help me God – a guy bouncing along like a whirlwind on bedsprings, little puffs of exhause rising between leaps as he jounced across the showroom, described a jittering figure-8 for the benefit of the crowd, and headed our way.

"Him and that crazy motorized-pogo-stick!" said Sparky, his voice loaded with admiration. "Ain't he the craziest?"

The "craziest" bobbled up and stepped off the mechanized stick. Casually, he dug a huge brush from his pocket and unhooked the

CONTINUED ON PAGE 62

The crowd of onlookers suddenly split wide open. "Von Dutch and that goofy motorized pogo-stick!" said Sparky with admiration. – "Ain't he the craziest?"

JUNE, 1956 25

Dutch's engineering drawings of the infamous motorized "Two-Stroke Pogo Stick" appeared in the letters section of the November 1955 issue of *Hot Rod* (left) and also advertised "two-stroke Roller Skate" [sic] and mentioned a "turbine yoyo." The letter added to Dutch's zany reputation and served notice that he was more than a pinstriper. In *Car Craft* the following June (above), a pogo-riding Dutch appeared in a story written and illustrated by Carl McKohler.

Dutch's first documented experience with ordnance (though a picture in the John Lawlor *1001 Custom and Rod Ideas* story "I Found Von Dutch" shows that Dutch spelled it "ordinance"—another of Dutch's pranks?) came when he was just 13. He found a .50-caliber Browning machine gun in a trench near Slauson Avenue and Alameda Street. The newspaper covering the finding reported that the young Kenneth Howard, who "has a boyish interest in things military, in fact, is quite a collector of war trophies," swapped the gun for a German saw-tooth bayonet.

While an obvious fan of firearms, there's very little hard evidence of Dutch's early gunsmith artistry. He was, however, a talented engraver and some examples of his gun engraving, reworking, and actual creation begin to appear in the mid-1960s. Indeed, there appeared in the August 1966 issue of *Guns* magazine, and also cited in *The Art of Von Dutch*, a letter in which he stated that he was "a working gunsmith." In the letter, he went on to talk about gun legislation, taxes, and other firearms issues in quite lucid terms. The man was also articulate. By 1971, his business card with his Tempe, Arizona, address advertised "Complete gunsmithing."

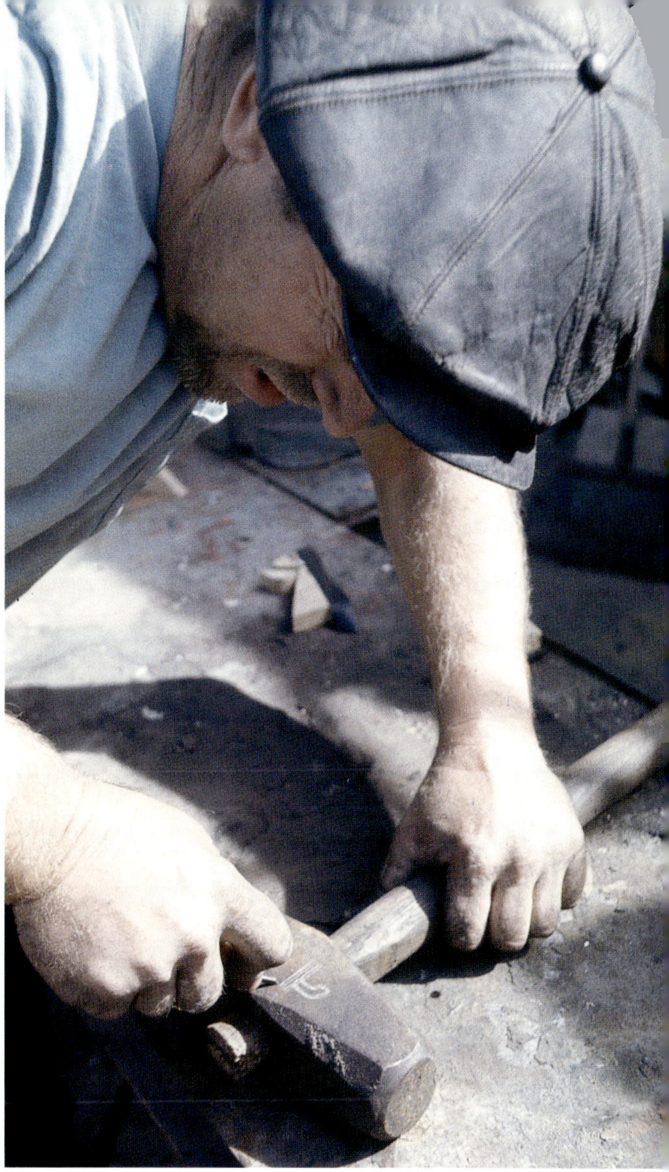

Dutch's fascination with Egyptian motifs has been well-documented in relation to the origins of his trademark "flying eyeball" (also see page 81). Here, he applies a decidedly Egyptian-looking glyph to a sledgehammer. Nolan reported that Dutch, a skilled engraver, used tiny handheld tools.

Perhaps one of Dutch's most interesting firearms was a replica of a .50-caliber U.S. Navy Elgin Cutlass Pistol (both gun and a knife) made for Jim Brucker. Dutch apparently fabricated everything: bored the barrel, carved the blade, and fashioned and engraved the wooden stock. There was also a nice scratch-built .357-caliber pistol that he supposedly made in two days while residing in Calabasas.

As Richard's photographs show, Dutch could not only make a gun, he could make a box to hold it—a weathered and lined presentation case complete with hand-fashioned tools and shell repository.

Dutch embellished his guns with engine turning, also known as a *damascening*, a term used to describe a decoration typically used in swords

Dutch's other and more obvious cultural admiration was for Germany. Richard muses that at first he believed Dutch tolerated his presence in Calabasas only because he had a suitably Teutonic name. The lathe inside the bus exhibits pinstriping, machine-turning, and a German legend which loosely translates to read: "Von Dutch's Weapon and Motorcycle Factory of Calabasas."

Dutch's first documented experience with ordnance came when he was just 13 and found a .50-caliber Browning machine gun in a trench near Slauson Avenue and Alameda Streets. It made the local paper. The handgun whose fabrication Richard documented was allegedly commissioned by Nolan as a souvenir of his time with Dutch (i.e., "Make me a machine that is a piece of art.").

but which Dutch applied to anything that didn't move–as well as a lot of things that did. Today, engine turning can be screened onto surfaces, but Dutch probably employed the age-old method of using the end of a steel bar dipped in grinding paste or fitted with a grinding disc. This was pushed onto the surface and twisted until it created the swirl pattern. Dutch applied the process to everything from triggers and barrels to knife blades, his louver press, and his sander. Jim Brucker recalled, "He had such a brilliant eye and great coordination he could do that stuff without a jig." Like engraving, engine turning took time and patience, which Dutch appeared to have in abundance when it came to his craft, though not so much so with people, who risked being ignored if they didn't have calloused, "working" hands.

Dutch seemed to accelerate his gun production in 1974 when he moved to Brucker's Movie World Cars of the Stars and Planes of Fame Museum in Buena Park, California. Before returning to Los Angeles from Arizona he was quoted in the July 8, 1973, edition of *The Arizona Republic* as saying, "Striping and guns are the only things I'm interested in."

At Movie World, where he worked alongside Ed "Big Daddy" Roth for a time, Dutch proved equally adept at illustrating his ideas for guns. The drawings are in both line and color and reveal yet another facet of Dutch the artist. Highly detailed and drawn freehand and without the aid of drafting equipment, they explore intricate mechanisms in the way that DaVinci might have explored a machine not previously invented.

One of the most interesting examples of Dutch's gunsmithing artistry was the 1976 crank-operated Krankenstein air rifle. Dutch's hand-lettered sign for the gun read, "A high power air rifle, it is not fool proof. The crank can bite back if not wound fully and the gun recoils when fired like a broken garage door spring."

The sign went on to say, "Took more time to make than anything in this case." Fashioned from wood, brass, and steel, the .22-caliber "Krank," as Dutch called it, apparently took 30 days to create. Of firing the gun, Suzanne Williams remembered, "Its weight was substantial but felt surprisingly unawkward. The tolerances were tight so it was difficult to ratchet. As I pulled the trigger I considered the possibility of a muzzle explosion. It went off with a hell of a kick. Man, that gun was sweet—what an exhilarating experience." In 2006, the Krankenstein sold at auction for $9,775.

Movie World closed in 1979 and moved to Santa Paula, about 100 miles north of Los Angeles. A couple of years later, around 1982, Dutch and his white GM bus joined them. He lived in a trailer alongside the warehouse—a huge defunct fruit-packing plant. It was about this time that

System Von Dutch came into being, and Dutch added the words to his business cards.

Accounts vary as to how many knives and guns Dutch fabricated. Jim Brucker would say 1,000 and 200 to 300, respectively; others far less.

Some have criticized Dutch for his use of ivory in making weapons, but he most likely used old piano keys and other discarded material, as he indicated on a label for a knife made for Cindy Rutherford in the early 1980s. Dutch rarely bought anything, always preferring to scavenge, trade, or otherwise use what was available. I remember my old friend, guitarist Jeff Beck, returning to England from a West Coast trip in the early 1980s with a knife. He'd gone to Dutch to have his '32 roadster striped. There was a knife available but it had someone else's name on it–Dutch merely engraved over it and handed it to Jeff.

Von Dutch's work with handguns accelerated after 1970, though he never fully gave up on painting like he told Tom Nolan he intended to do. By 1971, his business card, with his Tempe, Arizona, address, advertised "Complete gunsmithing."

Omega
Omega

To a craftsman, quality control is everything. Dutch sights down the barrel of the finished gun. In *The Rodder's Journal*, No. 31, Thom Taylor noted that Dutch was never known to actually machine his barrels. Nolan, on the other hand, reported that "Von Dutch hollowed lengths of iron with his tools." One way or the other, more fodder for the legend.

Dutch and a pair of detail photographs taken of his bus/machine shop. Nolan reported that Dutch had experienced difficulties with state regulators who discouraged his operation of a one-man shop. "The bus," Dutch noted, "is a movable vehicle, and their codes don't apply." Perhaps the Dymo label below was to discourage further meddling from the suits.

Second to quality control is presentation. Dutch was also a talented engraver, though, in typical fashion, he brought his own technique to the practice. "Engraving takes an enormous amount of strength," he told Nolan. "I don't like to do it for gun collectors. The only thing they give a damn about is that the engraving should be exactly the same on both sides of the barrel. It gets boring." As Richard's photographs show, Dutch could also make a handsome box to hold a gun–a weathered and lined presentation case complete with hand-fashioned tools and shell repository.

DYNAMIT NOBEL AKTIENGESELLSCHAFT NÜRNBERG
100
RANDZÜNDER
LANG
DESIGNED & MADE BY
Von Dutch
1970

HORSELESS CARRIAGE 1119

Chapter 4

Make Me a Machine

DESPITE BEING NOMINATED FOR TWO OSCARS, *The Reivers* was not a memorable movie, especially for Steve McQueen fans, who were expecting their action hero to reenact the previous year's hit *Bullitt* or the hugely popular 1963 movie *The Great Escape*. That masterful 1963 World War II film had allowed McQueen to establish Solar Automotive Products, and one of his first projects was *The Reivers*, based on William Faulkner's Pulitzer Prize–winning tale of the same name published in 1962, one month after the author's death.

Left: **Dutch and wife Sheila with the Winton Flyer. "In some ways," Richard says, "Dutch, Sheila, and their daughters were so 'suburban' I generally forgot that he was a living legend." Richard remembers Sheila as an "easygoing gal" who got on well with Richard's girlfriend at the time. He also wonders if anyone saw Dutch smile as much as he did on those four Sundays.**

Right: **Dutch fit the single-cylinder motor to this magnesium-alloy tricycle for his daughters, Lisa and Lorna. And to think nearly every kid you see riding a pedal-powered bicycle or tricycle today is wearing a helmet.**

In *The Reivers*, McQueen played Boon Hogganbeck, according to the tagline, "a lying, cheating, brawling and womanizing" plantation handyman and a reiver (an old Scots word for robber). Boon looks back 60 years to when he "borrowed" the plantation owner's then-novel car–a Winton Flyer–and set out with the owner's grandson Lucius (played by Mitch Vogel, later Jamie Hunter Cartwright in *Bonanza*) on a road trip 80 miles from Mississippi to Memphis. Traveling with them was Ned (played by Rupert Crosse), generally accepted to be Lucius' cousin. As the story progresses, Ned trades the bright yellow Winton for a racehorse, which Lucius must ride to win back the car. In doing so, he discovers the world of adults for all its foibles and redemption. Dutch had a walk-on bit in the film.

Of course, there was no such car as the Winton Flyer–it was a creation of Faulkner. However, The Winton Motor Carriage Company of Cleveland, Ohio, was a pioneering U.S. automaker, incorporated in 1897 by Scottish immigrant Alexander Winton, owner of the Winton Bicycle Company. The first American company to sell a motor car, Winton marketed automobiles to upscale consumers through the 1910s. Innovation and competition led to the end of Winton automobile production in 1924, but in 1912 the company had begun making marine and stationary gasoline and diesel engines, and in 1930 it became the Winton Engine Corporation, a subsidiary of General Motors.

So, did Dutch really build the Winton Flyer? Well, not from scratch he didn't. In William Nolan's 1972 book, *Steve McQueen: Star on Wheels, McQueen* is quoted: "I went to my old pal Von Dutch and I laid the problem on him. This cat can build anything from guns to tanks, and he agreed to supply our Winton. Made it all himself, from the frame up, out of old aluminum furnace sidings and God knows what else. When he showed it to us we all gasped. The thing was just beautiful–bright yellow, gleaming with polished brass–a perfect replica of a genuine 1905 Flyer, but with a modern power plant under the hood."

It appears that because McQueen's people were unable to find a real Winton they acquired a circa 1910–1912 E-M-F Flanders. Like Winton, the E-M-F Company was an early American automaker, its name gleaned from the

Right: **Though it's not the same louver press that fetched $11,500 at auction in 2006, it's possibly more stunning to behold, with its fearsome moniker and, well, louvers.**

initials of its three founders: Barney Everitt (a custom auto-body builder from Detroit), William Metzger (formerly of Cadillac), and Walter Flanders (Henry Ford's ex–production manager). E-M-F produced several models of its own design and contracted with Studebaker to sell them though their dealerships. They were notoriously poorly built, and detractors soon began quipping that E-M-F variously stood for "Every Morning Fix-it," "Every Mechanical Fault," and "Every Misfire." Unhappy with E-M-F's poor quality and lack of management, John M. Studebaker gained control of the assets and plant facilities in 1910. At a cost of $1 million, Studebaker paid mechanics to visit every dissatisfied owner and replace the defective parts in their vehicles.

Von Dutch's Winton was, in typical Dutch fashion, an amalgam of parts. Apparently, he discarded the back of the E-M-F and built a new seating area, according to McQueen and to Tom Nolan, quoting Dutch in *West*, out of aluminum furnace siding. For a powertrain, he utilized a British inline four-cylinder, possibly out of a Morris Minor. However, the intake was hand-fabricated and fitted with an old Amal motorcycle carburetor with a remote float bowl and a simple choke made out of brass–a neat bulge in the hood side accommodated it. The pedals were also fashioned out of brass, though the brake and clutch appear to operate a Chevy dual master cylinder. No doubt Dutch also crafted the brass Winton Flyer logo for the grille.

The hood sides and various other panels on the car were louvered, presumably by Dutch on his trusty louver press (which fetched $11,500 at the 2006 Brucker auction). The louvers were striped, as were the fenders and most of the body panels, in traditional black-on-yellow coach-type striping, although the Winton Flyer is reported to have been repainted and striped with plastic tape since its original construction.

On the car's oak firewall, a small brass plaque reads, "MFG in Calabasas by Von Dutch for Solar Productions 1968." The plaque is above and to the left of a "smoker" device that could lay down clouds of gray smoke by dripping oil into the number-one cylinder. Dutch also incorporated devices to aid starting (even though the car looked like it was started by a hand crank) and to cause shaking.

The
AWFUL
Tower

Winton Flyer
CALIFORNIA
HORSELESS CARRIAGE 1119

Winton Flyer
CALIFORNIA
HORSELESS CARRIAGE 1119

Winton Flyer
CALIFORNIA
HORSELESS CARRIAGE 1119

Winton Flyer
CALIFORNIA
HORSELESS CARRIAGE 1119

Winton Flyer
CALIFORNIA
HORSELESS CARRIAGE 1119

Winton Flyer
CALIFORNIA
HORSELESS CARRIAGE 1119

Winton Flyer
CALIFORNIA
HORSELESS CARRIAGE 1119

Winton Flyer
CALIFORNIA
HORSELESS CARRIAGE 1119

Von Dutch's Winton Flyer was, in typical Dutch fashion, an amalgam of parts, including a cannibalized E-M-F, aluminum furnace siding, an inline four-cylinder of British lineage, and an old Amal motorcycle carb. However, the intake and, no doubt, the Winton Flyer logo on the grille were hand-fabricated. Nolan breathlessly described the car as "a dream distilled."

Winton
Flyer

Winton

SAFETY FILM
2

Brass Era redux. The Winton Flyer was a device of William Faulkner, but Dutch brought it off the page, perfectly capturing the turn-of-the-twentieth-century juxtaposition of the industrial and organic worlds. In his 2006 book, no less an authority than Robert Williams points out, "A lot of people got satisfaction from looking forward to a machine age. [T]here was a hope that we could get away from the . . . foibles of being human beings. It was a cold philosophy." But Dutch, Williams said, "Could take the most pragmatic machine and put some wild pinstriping on it and relax it. It gives machinery a soul."

HORSELESS
CARRIAGE 1119

HORSELESS
CARRIAGE 1119

TEXACO
TEXACO
KODAK

Indian

ENC

Richard recalls that Steve McQueen's 1923 Indian (previous pages) which Dutch restored for McQueen, and the *Reivers* car were kept at the actor's Solar Automotive Products facility. His well-stocked sled/shed doubled as a paint booth for such projects.

Dutch takes his airbrush to a project (though not the Indian), sans respirator. Richard used a zoom to shoot the action from outside the shed, through the door (and a haze of red paint).

Steve McQueen kept his World War II–era American military halftrack at Dutch's property in Calabasas. "Dutch loved to run McQueen's halftrack around his 'ranch,'" Richard recalls. "Who wouldn't want to drive a roaring, unstoppable, multi-ton beast?"

U.S. ARMY 401503

Left: **The artist at rest. Dutch relaxes in the Calabasas ranch house, photos of family and Steve McQueen on the wall behind him. With the possible exception of the baseball bat leaning against the wall in the corner, the photograph seems to back up Richard's memories of a contented, regular guy–contrary to both contemporary and modern perceptions.** *Opposite:* **Dutch at the wheel of McQueen's halftrack with Kathy Morgan, Tom Nolan, and an unidentified child in the back.**

Chapter 5

The Man and the Moon

DUTCH'S WORK HAS INFLUENCED MANY ARTISTS OVER THE YEARS, but his early work had an immense impact on young painters such as Dean Jeffries and Ed Roth. Jeffries' work, in particular, bears a strong resemblance to experiments Dutch was making at the time. Occasionally, the two collaborated, most notably on *The Flamer*, George Sein's '32 Ford five-window coupe chosen in 2006 as one of the 75 most-influential Deuce hot rods of all time. It was unearthed with much of its original Dutch and Jeffries striping and painting still intact.

Left: **With "The Awful Tower" visible in the background, Von Dutch examines the moon globe he pinstriped for Richard–a keepsake the photographer still owns to this day. Richard explains that throughout 1969 and 1970, NASA's moon missions seemed to pervade the nightly news and that the moon seemed a time-appropriate subject for Dutch's revered skills.**

Dutch was not necessarily happy about the notoriety, adulation, or blatant copying he inspired, telling Eugene Jaderquist in *True's Automobile Yearbook 1957*, "All the guys who were painting cars in town came over to discover my secrets. They'd watch me like I had something up my bare arms. Then they'd begin to shift their way over to the work bench. They could see I didn't use any gimmicks while I worked, so they figured it was all in the brushes. Or in the paint. They swiped so many brushes I almost went broke replacing them."

Jaderquist went on to write, "These remarkable gifts–the accurate eye and the steady disciplined hand–are evident not only in car striping but in the paintings Dutch turns out. . . . Many, like the giant 12x12-foot mural that now decorates one wall of his apartment, are akin to the earlier works of Salvador Dali. They show the same clean, precise draftsmanship and unorthodox imagination. Others though more impressionistic leave little doubt that their creator has talent." The mural, just visible in some family photographs, is surrealist in form and shows an urn of what we suppose is wine tipping over from atop a pedestal.

Interesting that Jaderquist should reference Dutch's "precise draftsmanship." In the preceding paragraph he also referenced Dutch's refusal to practice using drafting tools in school, saying he didn't need them: "He could draw circles, curves and straight lines without any equipment at all. He did not complete the course." For evidence of this claim, one need look no further than Dutch's plans for the "Two-Stroke Pogo Stick" that appeared in the November 1955 issue of *Hot Rod* (see page 66).

Dutch's ability to achieve with freehand the sort of precise results associated with drafting tools is exemplified on the Rand McNally moon globe that Richard Koch took to Dutch to stripe on his last weekend of photographing him. Richard recalls that throughout 1969 and 1970, the *Apollo* space program seemed to pervade the nightly news. In fact, the nearly disastrous *Apollo 13* mission played out just weeks before the *West* article appeared, perhaps even as Dutch striped Richard's globe. What more appropriate souvenir to take away from his weekly visits with the man whom he had come to admire? Richard reasoned. When the moon was completed,

MARE
COGNITUM
MARE
HUMORUM
FRA MAURO
LEON
HILLS
SCHICKARD
HAINZEL
WILHELM
LONGOMONTANUS

Richard took it home and projected one of his 35-millimeter slides of Dutch's face onto the moon and photographed it, much as he had projected Dutch's striping from the Plymouth Duster onto Dutch himself.

To achieve the symmetry exhibited on the globe, Dutch adhered to the words he wrote to Ruth Levy Raymond: "To make an impossible straight line with a paint brush means [to me] overcoming with maximum concentration the fact that we are made out of meat, not steel—I used to concentrate and chant [in my mind] 'You are a machine' over & over, over, over, over. Once realized, the ability to pull a straight line in this world of incompetents gives one a great power! Over what I don't know but a great power!"

Striping the globe "Took Dutch no more than 60 minutes," Richard marvels. "And it's hard to see anyplace where the striping on one side differs from the other—and it was all done freehand." Dutch's tiny, precise signature measures only 13 millimeters in length, and Richard surmises that Dutch chose its location for its proximity to geographical features with suitably Germanic names.

Of course, Dutch more than once pulled a clever stunt when showing detractors how to "pull a perfectly straight line." He would weight his dagger brush with paint and let its own weight pull it down the surface.

In the *West* article, Tom Nolan quotes Dutch as saying, "I think I'm going to stop painting altogether soon." Of course, Dutch couldn't stop. He was compelled to create. In *The Art of Von Dutch*, his friend and customer Phil Lebow recalls, "I think one of the reasons Dutch was such a colossal alcoholic was because it was the only way he could come down off of his compulsion to work all the time. He didn't have much use for people because they just used up his time."

While Dutch is said to have preferred function over frivolity, he created numerous pieces of art and found the act of painting cathartic. Dutch's early influence was the surrealist school exemplified by Salvador Dali. The caption below a picture of Dutch striping an MGA in a 1950s issue of *Sports Motoring* read, "Known as the 'Salvador Dali' of auto decorators, Von Dutch of Hollywood is shown striping a new MG roadster fender. He uses his ear as a handy brush holder."

Dali comparisons and Dutch's mastery were further confirmed in the *Los Angeles Herald & Express* of June 7, 1958, which reported Dutch was "rapidly acquiring a reputation as the Salvador Dali of automotive painting." More than a quarter-century later, Henry Flood Robert Jr., writing in the Los Angeles Museum of Contemporary Arts' "Automobile and Culture" exhibit publication of 1984 stated, "Von Dutch was the abstractionist; Tommy the Greek the masterful straight liner."

No less an authority than Robert Williams, the father of the so-called "lowbrow" art movement, recalls in *The Hot Rod World of Robt. Williams*, "I saw Dutch in *Hot Rod* magazine in 1953, and it was an ignition point for me. He had a bigger effect on me—a far bigger effect on me—than Ed Roth did. . . . Von Dutch was up there with Salvador Dali in my mind. I immediately saw it—he was giving an abstract quality, a humanist quality, to hot rods. If you look at hot rods prior to his era, you'll see flame painting but it was very simplistic. Not only did Von Dutch do flame painting and pinstriping, he did biomorphic and anthropomorphic surrealist figures and that really got me."

Dutch's beloved surrealist movement was founded by French writer André Breton, who was influenced by the writings of Sigmund Freud. Breton published *Manifestoes of Surrealism* in 1924, and opened a studio for "surrealist research." Surrealism, according to its disciples, reunited conscious and unconscious realms of experience so completely that the dream and fantasy world was joined to the everyday rational world. In 1929, Dali said of the movement, "Surrealism is destructive, but it destroys only what it considers to be shackles limiting our vision." Dutch's vision was anything but shackled.

DO NOT CRUSH
FRAGILE
ONE RAND McNALLY
GLOBE
7074
Handle With Care

With Tom Nolan looking on, Dutch unwraps and assays Richard's Rand McNally moon globe for the first time. Says Richard: "When he opened the box his face lit up with a kind of goofy smile, and for a few seconds he held it in front of his face."

"Like all good pack rats, of which he was certainly one, Dutch had just the thing in his junkpile with which to hold it up while he striped it," Richard recalls. "It was a long metal rod with some kind of attachment on one end that seemed to have been put there in the distant past by Providence for just that use."

26

CHROME X FILM
28

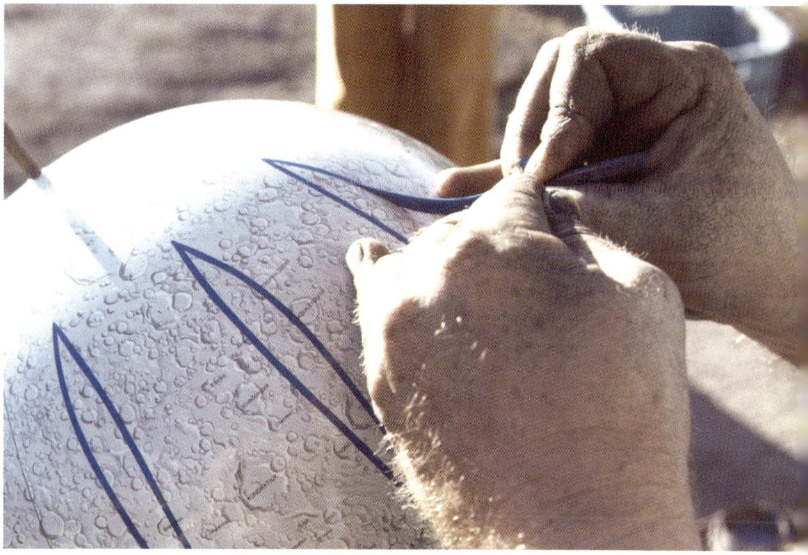

Richard: "When he first got the moon in his hands, Dutch studied it for quite a while. Then he started striping it." Richard still marvels at the symmetrical nature of the globe's pinstriping. "It's hard to see anyplace where the striping on one side differs from that on the other." Dutch more than once pulled a clever stunt when showing detractors how to "pull a perfectly straight line," weighting his dagger brush with paint and allowing its own weight to pull it down the surface.

Richard notes that Dutch striped the globe in two stages and that the whole process took no more than an hour. "One of my shots shows the first time his brush touched it, and another shows the moment he flicked the brush up off his logo in a kind of final flourish," he says.

With his three-legged striping table by his side, Dutch puts his finishing touches on the globe. In the background, behind Dutch's table and baking in the SoCal sun, is what appears to be a Ferrari 250 GT with its headlight rings removed. Richard recalls that the girl looking on was one of Dutch's two daughters.

Dutch signs the globe. "After he finished the striping, he studied it some more and then, using that long-bristled striping brush, he painted his impossibly tiny signature on it," Richard says. "The entire signature is only 13 millimeters long, yet when it is blown up, it is totally legible, down to the apostrophe in front of the '70.

"I can't help but wonder if Dutch chose the particular spot on the globe for his signature that he did because of the preponderance of German names around it. I also can't help but wonder what fantasies might have been going through his mind when he thought of his name there among those of the great German astronomers, even if only on a small replica of the moon."

Author Biographies

Richard Karl Koch began his professional life as an architect while still an award-winning student of architecture. He abruptly switched careers and spent 12 years as a successful architectural and lifestyle photographer. His clients were the top architectural firms in central Illinois and, after moving to California in 1966, most of the top firms in Southern California. His work appeared in architectural magazines and photo exhibits, and his lifestyle stories were published in numerous magazines, including major cover stories in *West* and *Home*, the Sunday supplements to the *Los Angeles Times*.

While shooting a story about the Santa Monica Pier for the original *Westways* magazine, he decided to create the West Wind Sailing Club and School in the summer of 1975. For the next five years he was involved with West Wind and as an officer of the Pier Promotion Association. He also ran (unsuccessfully) for Santa Monica City Council on his own "Save the Pier" platform. Following a major storm that washed away 300 feet of the pier–and, with it, his sailboat fleet and business premises–Richard began manufacturing and marketing a conditioning device of his own design for windsurfers; it was used by the U.S. Olympic windsurfing team and helped one member win the silver medal at the 1984 Summer Olympics. Shortly thereafter, Richard became involved in two lengthy real estate acquisitions and in 1992 he returned to the less physically demanding practice of studio photography, specializing in product advertising. Medical problems forced him to retire in 1996.

Presently he is struggling to get cranky contractors to complete the small unfinished house he purchased in 2005 near the foot of Bartlett Mountain in the high desert community of Joshua Tree, California, also known as "The New Bohemia" for its high population of artists of all kinds. His plans, if he ever completes the house, are to involve himself in his studio and in the activities of the Morongo Basin Cultural Arts Council, and to photograph the wildlife population on his five-acre front yard overlooking Joshua Tree and the mountains that surround the Joshua Tree National Monument.

Richard shares his life with his much-loved and faithful companion, Buddy the Lightning Dog. They live in Long Beach, California.

Courtesy Nathan Claydon

Tony Thacker is the executive director of the Wally Parks National Hot Rod Association Motorsports Museum and the author and co-author of several books on hot rods and customs, including *Hot Rods by Ed "Big Daddy" Roth* (Motorbooks, 1995), *SO-CAL Speed Shop* (Motorbooks, 2005), and *'32 Ford Deuce* (Motorbooks, 2007).

Selected Bibliography

Baldwin, Jack. "'Are You Von Dutch the Car Striper?' 'That's Me!'" *Car Craft*, Vol. 3, No. 9, February 1956.

Caldwell, Bruce, ed. *The Best of Hot Rod*. Los Angeles, Calif.: Petersen Publishing, 1981.

D'Olivo, Bob. "Camel's Hair Capers: The Fine Line between the Mediocre and the Mighty." *Hot Rod*, Vol. 8, No. 2, February 1955.

D'Olivo, Bob. "Camel's Hair Capers." *Hot Rod*, Vol. 8, No. 2, February 1955.

D'Olivo, Bob and Eric Rickman. "Motor Review '55." *Car Craft*, Vol. 3, No. 8, January 1956.

Ganahl, Pat. *Von Dutch: The Art, The Myth, The Legend*. North Branch, Minn.: CarTech, Inc., 2005.

Greene, Bob. "Trick with a Tank." *Hot Rod*, Vol. 8, No. 2, February 1955.

Jaderquist, Eugene. "Von Dutch: Flames, Stripes and Eyeballs." *True's Automobile Yearbook 1957*, No. 6.

Kahan, Aaron, Douglas Nason, Al Quatttrocchi, and Jeff Smith. *The Art of Von Dutch*. Los Angeles: Tornado Design, LLC, 2006.

Lawlor, John. "'I Found Von Dutch' and Lived." *1001 Custom and Rod Ideas*, Spring 1967.

McKohler, Carl. "The Legend of Ledhed McSlabb." *Car Craft*, Vol. 4, No. 2, June 1956.

Moon, Bob. "The Latest in Street Coupes." *Rod & Custom*, Vol. 1, No. 10, February 1954.

Nolan, Tom and Richard Koch. "Make Me a Machine That Will Be a Piece of Art." *West* (*Los Angeles Times*), May 24, 1970.

Nolan, William F. Steve McQueen: *Star on Wheels*. New York: Berkley, 1972.

"Restyled Wagons." *Car Craft*, Vol. 3, No. 12, April 1956.

Rickman, Eric. "Custom 'Rodster.'" *Hot Rod*, Vol. 6, No. 6, June 1953.

Stecyk, C. R. and Bolton Colburn. *Kustom Kulture: Von Dutch, Ed "Big Daddy" Roth, Robert Williams & Others*. Laguna Beach, Calif.: Laguna Art Museum and San Francisco: Last Gasp, 1993.

"Striping a Work of Art." *Rod & Custom*, Vol. 2, No. 11, March 1955.

Taylor, Thom and Richard Koch. "Von Dutch on a Sunday Afternoon." *The Rodder's Journal*, No. 31, 2006.

"Torrid '29 Tudor." *Hot Rod*, Vol. 7, No. 12, December 1954.

Von Dutch. "Post Entry." *Hot Rod*, Vol. 9, No. 11, November 1955.

Williams, Robert with Mike LaVella. *The Hot Rod World of Robt. Williams*. St. Paul, Minn. Motorbooks, 2006.

Index